ታሪከ ቁፅሪ

THE NUMBER STORY

SMALL BOOK ONE

ENGLISH - TIGRINYA

Numbers Teach Children Their Number Names

written and illustrated by

MISS ANNA

Early Reader Edition of *The Number Story 1*
Bronze Medal Winner, 2016 Wishing Shelf Book Award

Library of Congress Control Number: 2018902040

Names: Miss Anna, author.
Title: Number story : numbers teach children their number names / Miss Anna.
Description: Portland, OR: Lumpy Publishing, 2018.
Identifiers: ISBN 978-1-949320-17-6 | LCCN 2018902040
Summary: The pictures and rhymes present stories which introduce numbers 0-10.
Subjects: LCSH Numeration—English—Tigrinya--Pictorial works--Juvenile literature. | BISAC JUVENILE NONFICTION /
Languages: English—Tigrinya
Classification: LCC QA141.3 .M57 2018 | DDC 513—dc23

Publisher: Lumpy Publishing
Website: www.missannabooks.com
Email: missanna@missannabooks.com

Paperback: ISBN 978-1-949320-17-6
Printed in the U.S.A.   1 3 5 7 9 10 8 6 4 2

Want to learn our number names?

ናይ ቁፅርታት ኣስማት ከመሃሩ ይደልዩ ዶ?

It is very easy and a lot of fun!

ኣዝዩ ቀሊልን መዝናነይን እዩ!

Say-along our little jingle

ንእሽቶ ታሪክና ምሳና ይድረፉ!

starting from Number One!

ካብ ቁፅሪ ሓደ ንጀምር!

1

 looks like my one finger.

ሓደ

ን ሓንቲ ኣፃብዕትና ትመስል።

ONE!
ሓደ!

2

TWO trails a tail.

ክልተ

ጭራ መኽታል::

A TAIL! ရှ!

3

THREE has bumps.

ሰለስተ

ሓባጥ ጎባጥ አለዎ::

BUMPY! ሐባጥ ጎባጥ!

4

FOUR   carries a sail.

አርባዕተ

ናይ ጀልባ ጨርቂ ይሽከም።

A SAIL!
ናይ ጀልባ ጨርቂ!
ናይ መርከብ ሽራ!

5

FIVE   is a racing track.

ሓሙሽተ

ናይ ሽቀድድም መኪና ሜዳ እዩ።

VROOM
Go!

6

S I X   curves like a snail.

ሽዱ ሽተ

ከም ዓረና ዝተጠዋወየ።።

A SNAIL!   A SNAIL!

7

SEVEN   has a sharp angle.

ሸውዓት

በሊሕ ኩርናዕ ኣለዎ::

BE CAREFUL! IT'S SHARP!

ተጠንቀቅ! በሊሕ እዩ!

8

EIGHT   is rollercoaster rails.

ሸሞንተ

ሮለርኮስተር እያ።

የ !
እስዪ!
YIPPEE!

9

NINE   is a bubble on a stick.

ትሽዓንተ

ናይ ዓፍራ ኩዕሶ ምስ ሽንጣር::

A BUBBLE!

ናይ ባፍራ ኩዕሶ!

10

TEN   is an eye of a whale.

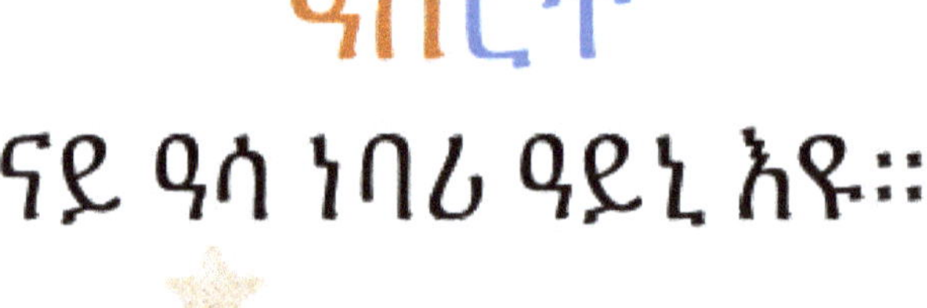

ዓሰርተ

ናይ ዓሳ ነባሪ ዓይኒ እዩ።

HELLO!

ከሰላም!

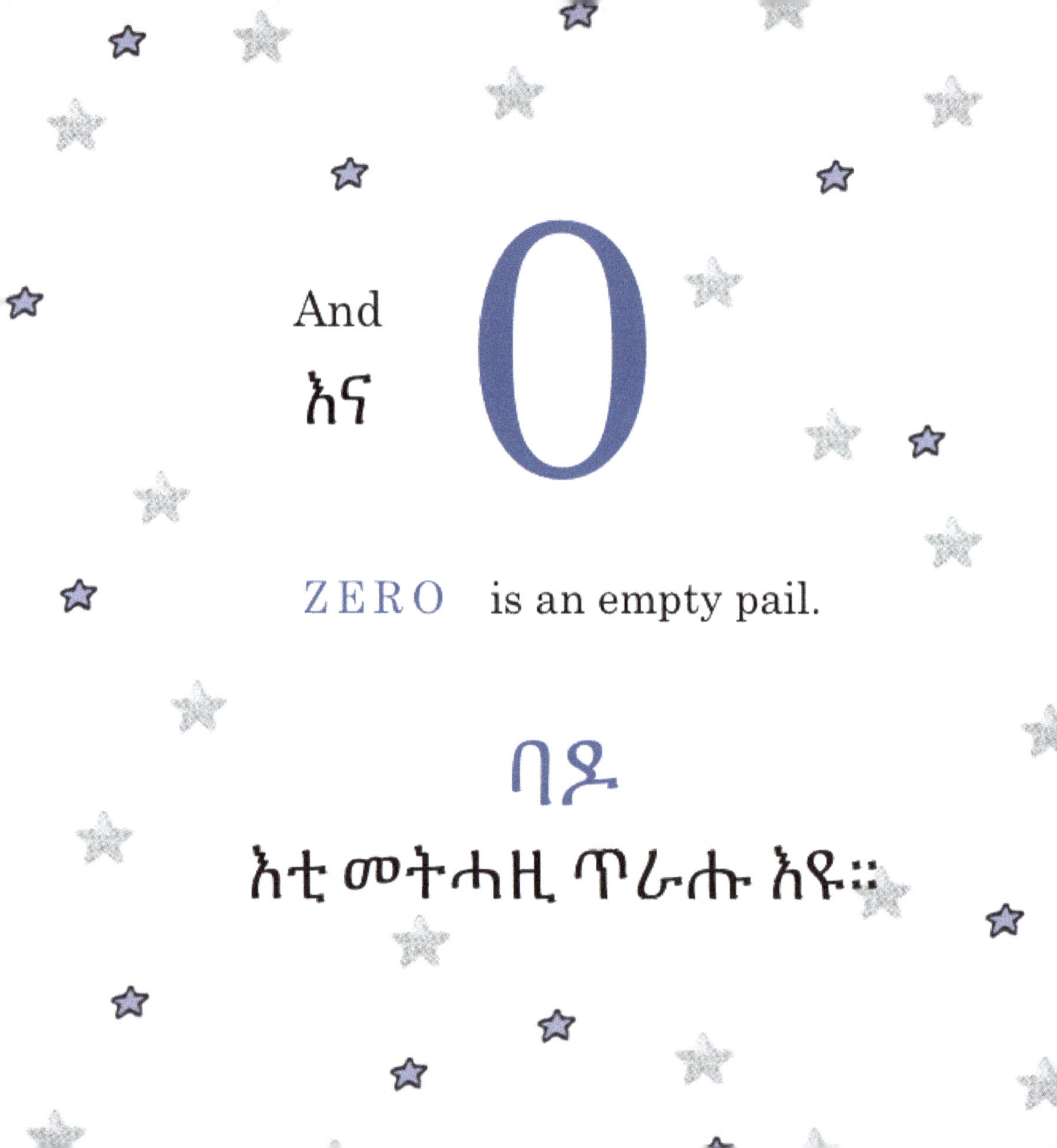

And

እና

0

ZERO is an empty pail.

ባዶ

እቲ መትሓዚ ጥራሑ እዩ።

IT'S
EMPTY!
ጥራሑ እዩ!

Thank you for playing with us today.

We had a lot of fun too!

ሎምዓንቲ ምሳና ብምፅዋትካ የቐንየልና።
ንሕና ዉን ብዙሕ ተፃዊትና!

We are your Number friends,
Zero to Ten,
Who will be here for you~
ንሕና ናይ ቁፅርታት መሓዙት ኢና
ካብ ባዶ ክሳብ ዓሰርተ፡፡
ኩልሻብ ኣብ ጎንኻ ክንከውን ኢና፡፡

Bye-bye now!
See you again soon!
ሕጂ ደሓን ኩን!
ኣብ ቀረባ እዋን ንራኸብ!

The Numbers are *SINGING* too!

To sing-a-long, look for Miss Anna Number Story
at your favorite music store like iTUNES.

MP3

Numbers 0-10
IDENTIFYING
& COUNTING

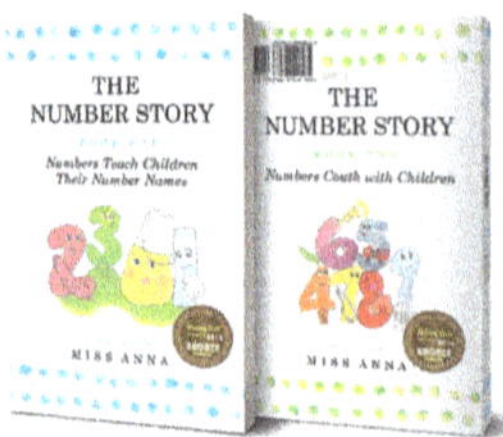

Number Story 1 & 2

isbn: 978-0-996216-48-7

Numbers 11-20
& Ordinals

first, second, third...

Number Story 3 & 4

isbn: 978-1-945977-01-5

Numbers 0-100
& Place Values

ones, tens, hundreds...

Number Story 5 & 6

isbn: 978-1-945977-06-0

About Clocks
& Telling Time

hours, minutes, seconds

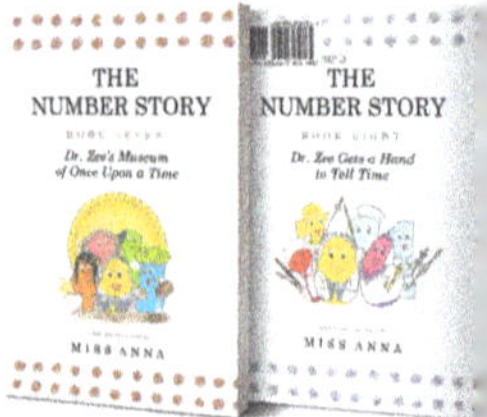

Number Story 7 & 8

isbn: 978-1-949320-40-4

For more Miss Anna books to love,
visit us at

www.missannabooks.com

Numbers are working hard all over the world!
Come Travel the World with Us!

www.ingramcontent.com/pod-product-compliance
Lightning Source LLC
Chambersburg PA
CBHW040902070726
47599CB00035B/2272